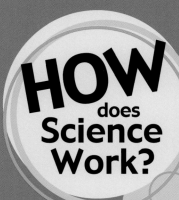
HOW does Science Work?

Exploring Electricity

Carol Ballard

PowerKiDS press.
New York

Published in 2008 by The Rosen Publishing Group, Inc.
29 East 21st Street, New York, NY 10010

First Edition

Commissioning Editor: Vicky Brooker
Editors: Laura Milne, Camilla Lloyd
Senior Design Manager: Rosamund Saunders
Design and artwork: Peta Phipps
Commissioned photography: Philip Wilkins
Consultant: Dr Peter Burrows
Series Consultant: Sally Hewitt

Library of Congress Cataloging-in-Publication Data

Ballard, Carol.
  Exploring electricity / Carol Ballard.
    p. cm. — (How does science work?)
  Includes index.
  ISBN 978-1-4042-4281-4 (library binding)
  1. Electricity—Juvenile literature. I. Title.

QC527.2.B375 2008
537—dc22
                                2007032229

Manufactured in China

Acknowledgements:

Cover photograph: Lightning bolt against night sky, Ivar Mjell/Getty Images

Photo credits: Betsie Van Der Meer/ Getty Images 4, Juan Silva/Getty Images 5, David Hiser/Getty Images 6, Alex Bartel/Science Photo Library 7, Peter Hendrie/Getty Images 8, Roger Tully/Getty Images 9, Claudia Kunin/Corbis 11, Luis Veiga/Getty Images 15, Andrew Lambert Photography/ Science Photo Library 16, Chris Knapton/Science Photo Library 20, Betsie Van Der Meer/Getty Images 22, oote boe/Alamy 24, LWA-Dann Tardif/Corbis 25, Alan Sirulnikoff/Science Photo Library 26, Tim Graham/Alamy 27, Ivar Mjell/Getty Images 28.

The author and publisher would like to thank the models Matt Barson and Alex Babatola, and Moorfield School for the loan of equipment.

Contents

Words in **bold** can be found in the glossary on p.30

Electricity

We use electricity every day in so many different ways that it is difficult to imagine life without it.

Heating systems keep our buildings warm and heat water for us to wash ourselves. Ovens and microwaves heat our food and drinks. Many smaller household **appliances**, such as vacuum cleaners and irons, also use electricity.

Fridges and freezers use electricity to keep our food cold and fresh. →

Without electricity, we would not be able to watch television or use a computer. Life-saving equipment in hospitals needs electricity to work. Street lights and ordinary light bulbs in our homes all use electricity.

Factories such as this one need electricity to make the machines work.

Making electricity

Anything that provides energy is called an energy source. This energy can be changed into electricity at a power station. Some power stations burn **fossil fuels**, such as coal, oil, and gas, for energy. These are made from the remains of plants and animals that died millions of years ago, buried deep under the ground.

Fossil fuels will not last forever. Eventually, they will run out and we cannot make any more of them. Also, using fossil fuels releases chemicals into the environment, which causes pollution.

↑ **Coal is found deep underground.**

Some power stations use **renewable energy** sources that will never run out, such as solar and wind energy.

Some power stations use **nuclear energy**, which is the energy locked inside **atoms**. This will last for a very long time. However, using nuclear energy makes **nuclear waste**, which can be dangerous if we do not get rid of it safely.

Using renewable energy, such as the wind, is better for the enviroment.

Electricity's journey

At the power station, energy from the fuel or other energy source is changed into electricity. Once it is made, the electricity travels to where it is needed.

People at control centers check where electricity is being made and used, and that there is enough of it. The power station that made your home's electricity may be many miles away. **Power cables** carry the electricity from the power stations.

Electricity is made at power stations like this one. →

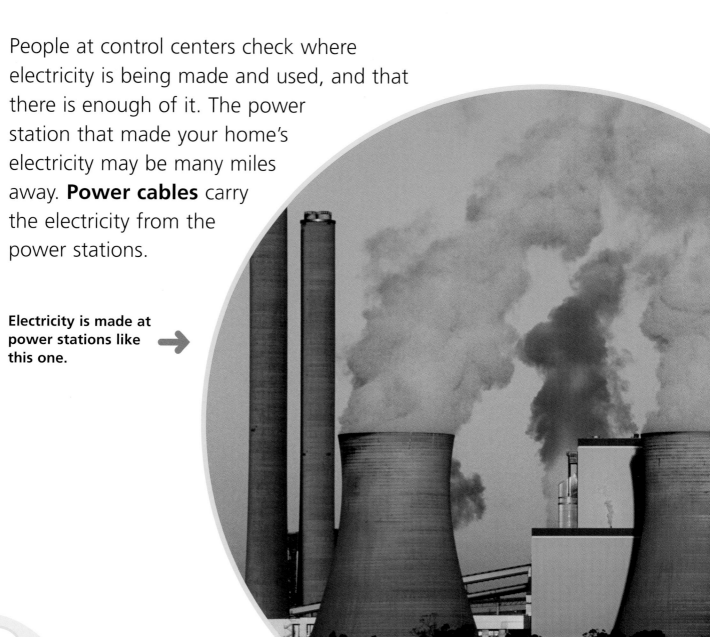

At this control center, people check how much electricity is being made and used. →

The electricity in the power cables is very powerful. This means that the cables are very dangerous. They are supported high above the ground by **pylons**.

In towns, transformer stations receive the powerful electricity and make it weaker. Underground cables carry electricity from transformer stations to buildings, such as houses, schools, and shops. When the electricity enters a building, it is the correct strength for the appliances to use.

Batteries

Batteries store energy. Small appliances get the energy they need from batteries. Solar batteries collect energy from sunlight.

When an appliance is switched on, the energy stored inside the battery is turned into the electrical energy the appliance uses.

This alarm clock uses a small battery.

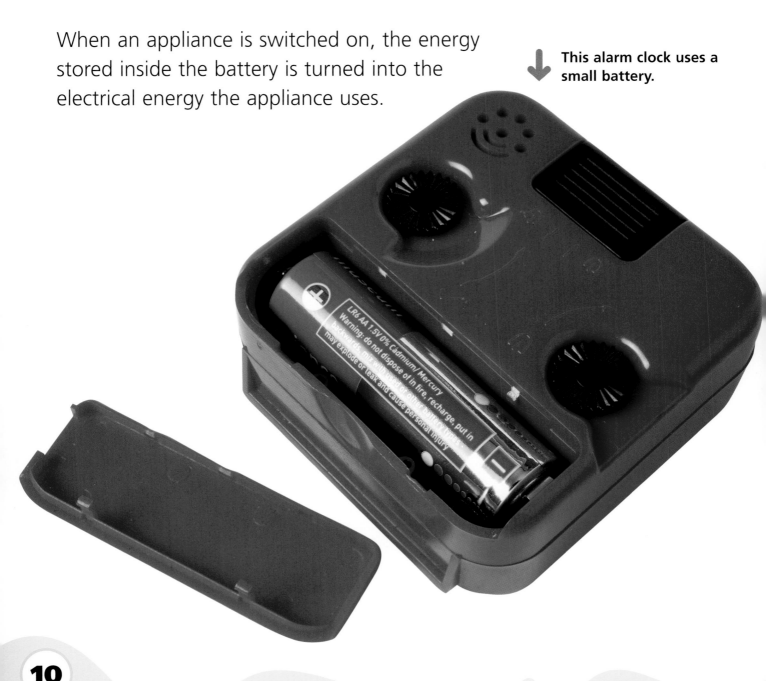

A battery's strength is measured in **volts** (V). The higher the voltage, the stronger it is.

Each end of a battery is called a **terminal**. One is positive (+) and the other is negative (-). If you are using more than one battery, you must connect the positive (+) end of one to the negative (-) end of the other.

Some batteries are tiny, such as those used in this remote control. The remote control is used to make the toy truck move.

Simple circuits

Bulbs and buzzers need electricity to work.
To get this, they have to be connected to a battery
by wires. Connecting them together makes a
circuit. Parts of a circuit are called **components**.

Electricity moves around the circuit like water
flowing through pipes. This is called an **electric
current**. It needs to be able to keep flowing
around and around in one direction.

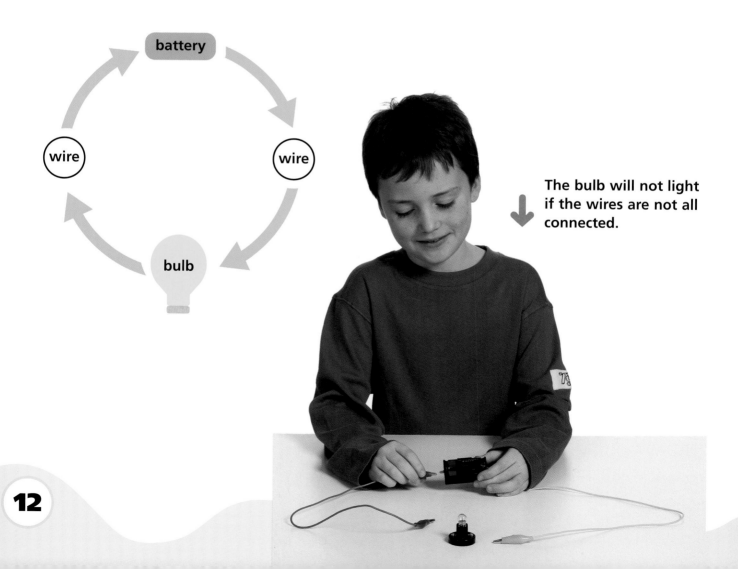

The bulb will not light
if the wires are not all
connected.

TRY THIS! Make a circuit to light a bulb

1   Gather a battery, two wires, and a bulb.

2   Attach one end of one wire to the positive (+) terminal of the battery with clips, adhesive putty, or tape.

3   Attach the other end of the wire to one side of the bulb.

4   Attach one end of the other wire to the negative (-) terminal of the battery.

5   Attach the other end of the wire to the other side of the bulb.

If your wires are all connected correctly, you should find that the bulb lights. The circuit is complete and the electricity flows all the way around it.

Adding more to a circuit

In the past, Christmas lights used to be in a **series circuit**, which meant that the current had to go through all the bulbs to light up. If one broke, then the electricity could not flow to the next one and none of them would light up.

Modern Christmas lights are wired in parallel so that if one goes out the rest stay on.

TRY THIS! Adding bulbs

1. Make a simple circuit as seen on page 13.

2. Look carefully to see how brightly the bulb is lit.

3. Detach the wire from one side of the bulb and attach it to another bulb.

4. Use a third wire to link the bulbs together. Look carefully to see how brightly the bulb is lit.

You should find that a single bulb is brighter than two bulbs. This is a series circuit. The current flows through both bulbs. This makes it harder for the current to get around the circuit.

A **parallel circuit** works by attaching separate wires to each side of the bulbs and then attaching all the free ends to the battery. The bulbs will both be as bright as the single bulb, because the current has a separate route to each bulb and does not have to go through the other bulb.

Conductors and insulators

Electricity can travel through some materials but not others. Materials that electricity can travel through are called **electrical conductors**. Materials that electricity cannot travel through are called **electrical insulators**.

Most plastics are good insulators. This is why metal wires are covered in plastic. The plastic makes the wires safe to touch, because the electricity cannot travel through the plastic.

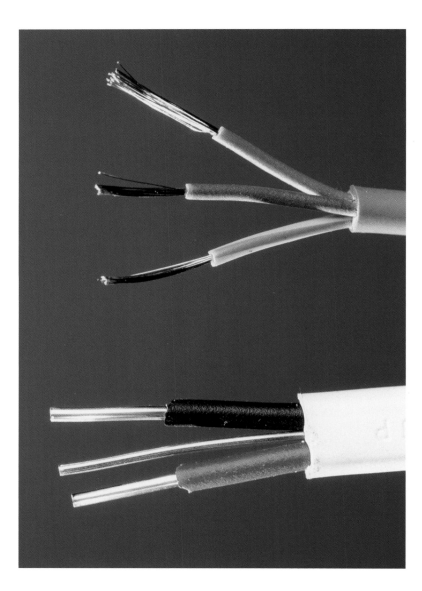

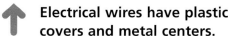 Electrical wires have plastic covers and metal centers.

TRY THIS! Test materials as conductors

1 You will need a battery, three wires, a bulb, and some small objects, such as a plastic straw, a metal key, and a rubber band.

2 Attach one wire to the positive (+) terminal of the battery and one side of the bulb.

3 Attach one end of a second wire to the negative (-) terminal of the battery.

4 Attach one end of the third wire to the other side of the bulb. You have a circuit with a gap.

5 Close the gap in the circuit by connecting the wires to each object in turn. Write down which objects let the bulb light.

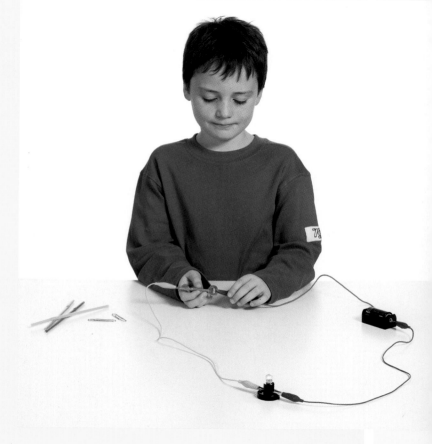

You should find that a bulb in a circuit only lights when metal objects are put in the gap. This is because electricity can flow through metal but very few other materials.

Switches

A **switch** is used to turn things on and off. Many things we use have switches. To turn something off, a switch makes a gap in the circuit—the electric current cannot flow across the gap. To turn something on, the switch fills the gap in the circuit—the circuit is complete and so the electric current can flow.

We use switches to turn things on and off. →

TRY THIS! Make a switch

1   You will need a battery, a bulb, three wires, a piece of card, two metal paper fasteners, and a metal paper clip.

2   Make a circuit with a gap as seen on page 17.

3   Attach one loose wire to each paper fastener, making sure the metal touches the fastener.

4   Slip the paper clip onto one of the fasteners.

5   Push one fastener through the card and open the points flat at the back.

6   Repeat step 4, making sure that the distance between the two fasteners is less than the length of the paper clip.

7   Swing the paper clip around so that it touches both fasteners at the same time, then swing it away again. Watch your bulb as you do this.

You should find that the bulb lights when the clip touches both fasteners. This is because the clip fills the gap in the circuit. The bulb should be off when the clip is only touching one fastener. This is because there is a gap in the circuit.

Resistors

Something that makes it hard for electricity to flow is called a **resistor**.

Wires are resistors. When electricity flows through them, they get hot. Electric fires contain a tight coil of wire. When electricity flows through this, it gets so hot that it glows bright red and gives off heat.

↑ A tiny coil of wire in a light bulb glows when electricity flows through it.

TRY THIS! Testing wires

1 Make a simple circuit as seen on page 13.

2 Look carefully to see how brightly your bulb is lit.

3 Now change one of your wires for a much longer wire.

4 Look at your bulb again.

You should find that the longer the wire, the dimmer the bulb. This is because the wire acts as a resistor—the longer the wire, the harder it is for the electricity to flow through it.

Some resistors can be adjusted to change the amount of electricity that flows. Resistors in volume controls turn the sound up or down on stereos. Resistors in dimmer switches control the brightness of light bulbs.

Electricity around the house

Electricity flows through a box called a **meter**, which measures how much electricity is flowing through. Electricity is measured in units called **watts**.

For safety, the electricity then flows through a **fuse box**. This contains **fuses** or **circuit breakers**. These will cut off the supply of electricity if there is a fault.

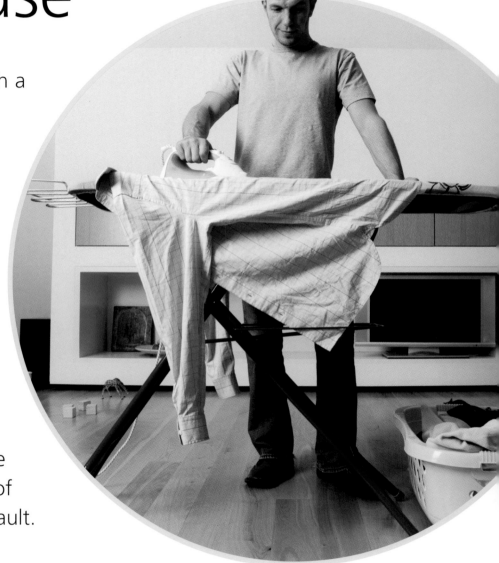

The electricity that arrives in your home can be used by household appliances such as irons.

Electricity flows through cables to reach every room in a house. The cables are usually hidden inside walls. One set of cables can carry electricity to the ceiling and wall lights.

Another set of cables carries electricity to the sockets in the walls. The sets of cables make circuits that we control with switches.

Cables have narrow wires inside. One wire is there to carry electricity safely into the ground if there is a fault. This is called an earth wire.

Which appliances in your room do you think will use the most electricity?

Household Electrical Appliance	Power (watts per hour)
Iron	1,200
Radio	40
Computer	250
Hairdryer	1,875
Lamp	125
Microwave	1,500
Toaster	75
Television	110
Video Game	20
Fridge/Freezer	800

Saving electricity

We use electrical appliances for many different things such as heating, light, cleaning, and entertainment. Different appliances use different amounts of electricity.

We use up a lot of electricity, but there are ways to save energy every day. Some electricity-saving inventions help us cut back on the amount of electricity that we use.

This energy-saving light bulb uses less electricity than a normal light bulb.

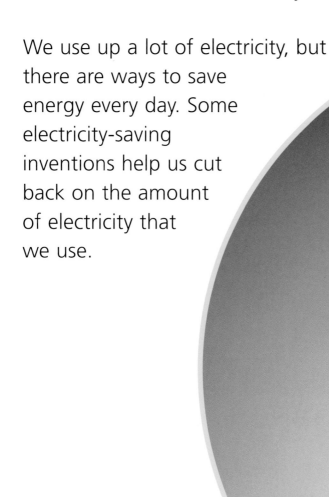

Every time we switch on an electrical appliance, energy is used. If the appliance is left on when we do not need it, energy is wasted. Not only does this cost money, it is also bad for the environment.

It makes sense to save energy whenever we can. This means remembering to turn off lights when a room is empty, not to leave appliances on "standby," and not to heat more water than we need for a hot drink.

Make sure that you turn off the television when you have finished watching it. →

Safety and electricity

Electricity is dangerous and can hurt you. Electricity is very powerful so you should use it wisely. Many appliances use electricity. They are plugged into a wall socket. Electricity travels through the socket into the appliance.

Never put anything except a plug into a socket or you could get an electric shock.

Never connect too many appliances to one socket, and do not touch frayed wires since these can be dangerous.

↑ Electrical sockets are dangerous if they are used incorrectly.

Never play with electricity

Electrical appliances can sometimes get too hot. Tell an adult and switch an appliance off immediately if you see smoke or smell burning.

Electricity can travel through water, too. Always dry your hands before switching on anything electrical.

Batteries can contain dangerous chemicals. Never try to take a battery apart or use one that is damaged or leaking.

Never play near overhead cables such as these by the railroad line. These cables carry strong electrical currents and can give you a shock or even kill you.

Static electricity

Static electricity is a type of electricity that does not flow. Most things around us contain tiny bits of electricity called **charges**.

These charges can be positive (+) or negative (-). Positive and negative charges attract each other. Positive charges move away from other positive charges. Negative charges move away from other negative charges. Usually there are the same number of positive and negative charges, so they just cancel each other out.

↑ **Static electricity causes a lightning flash.**

In a thunderstorm, electrical charges jump between clouds or between the clouds and the ground. This is called a flash of lightning.

TRY THIS! Make static electricity

1   You will need a balloon and a woolen sweater.

2   Blow the balloon up and knot the end.

3   Rub the balloon up and down on your sweater.

4   Hold it against a wall —it should stick!

This is because the balloon picks up extra negative charges from the wool. The balloon sticks to the wall, because the negative charges are attracted to the positive charges in the wall.

Wow!

A lightning flash happens when electricity builds up in the clouds!

Glossary

appliance something that uses electricity to work

atoms the tiniest parts of anything that can exist

circuit a complete path that electricity can flow around

circuit breakers things used to stop the flow of electricity

charges unbalanced (too much or too little) electrical energy

components things that make part of a circuit

electrical conductors materials that electricity can travel through

electrical insulators materials that electricity cannot travel through

electric current electricity that is flowing around a circuit

fossil fuels fuels found on Earth that were made millions of years ago

fuses simple circuit breakers

fuse box fuses that stop too much electricity flowing in a circuit

meter something that measures the amount of electricity used

nuclear energy the energy locked inside atoms

nuclear waste waste made when nuclear energy is used

parallel circuit where each component has a separate link to the battery

power cables strong wires to carry electricity

pylons strong frames that hold the cables

renewable energy energy that will not run out

resistor something that reduces the flow of electricity

series circuit where each component is linked one after the other to the battery

static electricity electricity that does not flow

switch something that turns things on and off

terminal one end of a battery

volts the units for measuring the power of a battery

watts the units for measuring how much electricity is used

Further information

Books to read

Conductors and Insulators (My World of Science) by Angela Royston (Heinemann Library, 2003)

Electricity (Start-up Science) by Claire Llewellyn (Evans Brothers, 2004)

Electricity: Turn It On! (Science in Your Life) by Wendy Sadler (Raintree, 2005)

Shocking Electricity (Horrible Science) by Nick Arnold (Scholastic Hippo, 2000)

Web sites to visit

Web Sites

Due to the changing nature of Internet links, PowerKids Press has developed an online list of Web sites related to the subject of this book. This site is regularly updated. Please use this link to access this list: www.powerkidslinks.com/hdsw/elect

CD Roms to explore

Eyewitness Encyclopedia of Science, Global Software Publishing

I Love Science!, Global Software Publishing

WARNING: mixing batteries of different types or amounts in the same circuit is dangerous and may cause an explosion.

Some advice for using batteries in simple circuits:

• the safest batteries for this purpose are zinc/carbon or zinc/chloride ones

• avoid using rechargeable or alkaline manganese batteries—because they have low internal resistance, they get very hot very quickly

• do read and follow all manufacturers' instructions on battery packaging

Index